IMPACT OF ELECTRIC CARS

Understanding propulsion, storage systems, environmental footprint of electric cars

Domenic Mönnich

ABSTRACT

This paper provides information on how the storage systems and the drive-in electric vehicles work and what effects these vehicles have on the environment and everyday life in their entire life cycle. Electric cars are often praised for their climate-friendly operation. This paper therefore explains the environmental factors of electric cars and the impact of their production, use and recycling. In order to investigate the reasons for these effects, the mode of operation as well as the differences in motorisation are examined. Advantages of an electric drive compared to conventional combustion engines are also shown. Batteries in particular have a major influence on the life cycle assessment of electric cars. Therefore, lithium-ion batteries are discussed in detail before alternatives such as water and methanol fuel cells are dealt with in detail.
The aim of this thesis is to analyse the life cycle assessment of electric cars with the help of literature and to provide information on how this technology will be developed and established in the future.

CONTENT

1. INTRODUCTION

"We will not stop until every car on the road is electric" is one of the most famous quotes from Elon Musk, the founder of the electric car manufacturer Tesla. Because Tesla is not the only one to focus on electric cars, more and more large vehicle manufacturers are researching, developing and producing electrically powered cars. Fossil fuels are only available in limited quantities and thus also the fuel for our cars. In recent years, the trend has therefore increasingly moved towards electric mobility. There are more and more electric cars on the roads and many vehicle manufacturers agree that electrically powered models will play an increasingly important role in the future. Electric cars are considered to be particularly environmentally friendly, as they do not require fossil fuel, but can be powered by electricity. But just because an electric car itself does not emit any exhaust gases does not mean that it is 100% environmentally friendly. From production to recycling, electric cars have a great potential to do something good for the environment. However, when it comes to life cycle assessment, even these vehicles are not perfect. The production of batteries and electricity in many countries is associated with high emission values and various environmental burdens.

In order to get an overview of all these factors, this literature work examines the life cycle assessment of the entire life cycle of an electric car. In order to explain the differences to conventional combustion engines, the electric motor is first explained and then its advantages and disadvantages are described. Since the battery in an electric vehicle plays a major role in the life cycle assessment, the functions and the positive and negative aspects of lithium-ion batteries are also shown. Then alternatives to the vehicle battery are discussed. Finally, it is analysed what influence electric cars will have on our future and what could still prevent the boom in electric cars.

2. LIFE CYCLE ASSESSMENT

2.1 Vehicle production

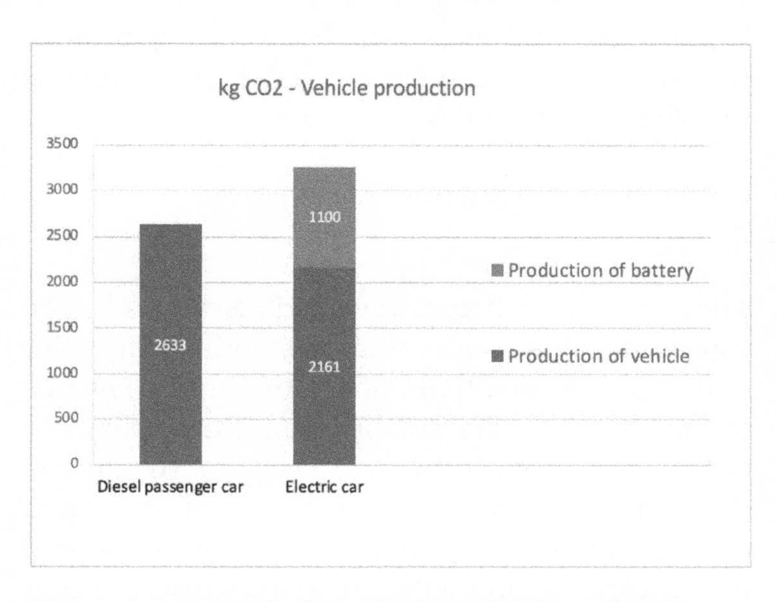

Figure 1, CO2- emissions during vehicle production

The construction of an electric vehicle requires a very high cost and raw material input. The main reason for this is battery production. Compared to vehicles with combustion engines, the production of electric vehicles is therefore more cost intensive. With the electric car, about 30% more environmentally harmful gases are emitted during production. The production of a 200kg transaction battery currently generates approx. 1,100kg of CO2 emissions. The transaction battery is the name given to the energy storage system consisting of several battery cells connected in parallel and in series. Half of the emissions are caused by the extraction and provision of the materials, the other half by the production and the associated energy consumption. Figure 1 shows the CO2 emissions from battery production and vehicle manufacture. (Cf.: Kairos - Institut für Wirkungsforschung, 2019, p. 4-5)

2.2 Battery production

The increasing demand for electric cars is leading to a growing mass production of accumulators. This requires large quantities of key materials. The materials of note here are lithium and cobalt. The faster the e-mobility sector develops, the greater the demand for these rare raw materials. In order to meet the growing demand, the required amount of lithium and cobalt could rise sharply over the next 30 years.
More than half of the world's cobalt comes from the Congo. Poor working conditions and the unstable political situation mean that the raw materials will

have to be sourced elsewhere. As a result, deep-sea mining is playing an increasingly decisive role in the extraction of cobalt, as around 83% of global cobalt reserves are located at the bottom of the oceans. The effects of this mining activity on the sensitive marine ecosystem have hardly been researched yet. (Cf.: Buchert, 2018)
With lithium, the problem is even greater. About 55% of all reserves are believed to be in the so-called "lithium triangle", consisting of Argentina, Chile and Bolivia. The production areas are among the driest in the world. However, the extraction of one ton of lithium salt requires two million litres of water. No consideration is given to the ecosystem in the mining area. (Cf.: Lauerer, 2018) Lithium production in these arid regions causes an enormous drop in the groundwater level. Unique natural sites, such as the salt lakes of the Andes, are destroyed in order to pump lithium salt out of the ground. No consideration is given to the population living there. The local flora and fauna also suffer from the mining of lithium. Chemical raw materials such as potash or caustic soda, which are needed for the further processing of lithium, cause enormous damage to the ecosystem. (Cf.: Thurn & Scherlofsky, 2019)

In addition to these two key raw materials, other minerals are also required for battery production, the extraction of which also poses problems:

- **Graphite** is an important component of the

anode of a traction battery. The largest producer of graphite at present is China. Due to a lack of safety precautions, enormous quantities of health-endangering fine dust are released during both mining and transport.

- In addition to cobalt and graphite, **nickel** is also used in most cathodes of transactic batteries. The Philippines, Indonesia, Russia and Canada are among the largest producers. During production, the air pollutant sulfur dioxide is released and thus pollutes the air quality in the affected regions and regions close to production.
- **Aluminium** is the main component of the housing and wiring. Only a small proportion is also used in the cathodes. Aluminium is extracted from bauxite, which is mainly mined in Australia. The extraction of aluminium by fused-salt electrolysis is very power-intensive. This process also produces the bauxite residue "red mud", which has to be stored and thus poses a risk to groundwater. (Cf.: Kairos - Institut für Wirkungsforschung, 2019, pp. 10-11)

The production of batteries is very energy-intensive. For cost reasons, almost all transaction batteries of German electric cars come from China or South Korea, especially since a large part of the electricity required for production comes from coal-fired power plants. However, these are very bad for the environment because of their relatively low efficiency but high CO_2 emissions. (Cf.: Thurn & Scherlofsky, 2019)

2.3 Utilization phase

The environmental aspects of noise, pollutant emissions and the CO_2 balance of the vehicle operation must be taken into account for electric cars during driving.

2.3.1 Noise

Road traffic is one of the biggest sources of noise. Urban traffic is particularly affected by low speeds. An electric motor is almost silent compared to a combustion engine. In the slow city traffic, this offers an enormous advantage in terms of noise. At higher speeds, however, this is less important, since the tyre road noise predominates.

2.3.2 Emissions Of Pollutants

Electric cars are locally emission-free. They therefore produce no local air pollutants. In polluted cities, electric cars could significantly minimize pollution. The amount of emitted substances is currently limited by the EURO 5 standard. The limit values apply to carbon monoxide, nitrogen oxides, hydrocarbons and fine dust particles. The stricter limit values for pollutants therefore practically do not affect electric vehicles.

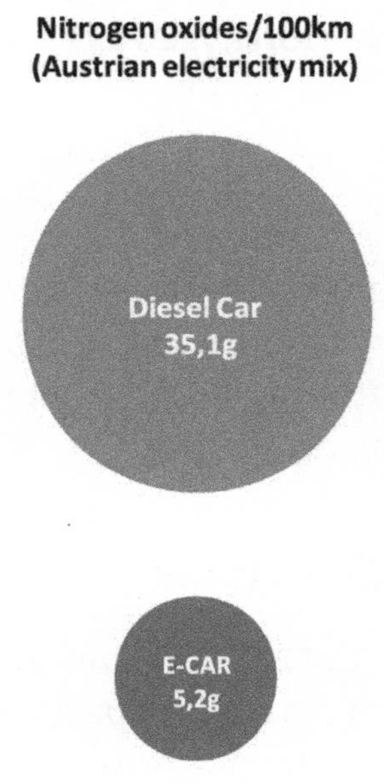

Figure 2, Nitrogen oxide emissions from diesel and electric cars

Although the electric car does not endanger the air quality on site, air pollutants are still produced during power generation. However, on average, the pollutants produced during electricity generation are much lower than those produced by burning fossil fuels in combustion engines. Shown here in figure 2.

Ina combustion engine, such pollution is not only caused by the engine, but also by abrasion of the clutch and brake linings. The design of the electric motor makes a clutch superfluous and there is no additional load. The regeneration of the electric motor also ensures that the brake pads are relieved and wear more slowly. The only point in which the electric car is equal to the combustion engine is tire wear. (Cf.: Karle, 2018, pp. 166-167)

2.3.3 Co2 Balance Of Vehicle Operation

Electric cars have a low energy consumption. The efficiency of the electric motor means that it requires only a third of the energy of a combustion engine over the same distance. If the electricity is generated from renewable energy sources, almost no carbon dioxide is produced. The CO2 emissions of an electric car during operation are much lower than those of a vehicle with a combustion engine. A diesel car of the compact class (5.2l diesel/100km) emits approx. 17,800g CO2 per 100 km driven, whereas an electric car (Austrian electricity mix) emits only 3,000g for the same distance. If the vehicle is powered only by green electricity, the amount of CO2 emitted is even reduced to only 400g. The less energy a country produces in a renewable and sustainable manner, the greater the associated CO2 emissions while driving.

The production of the energy source therefore plays a major role in the CO2 balance. The extraction of oil is much more expensive and environmentally harmful than the production of electricity. During oil production, disasters, conflicts and accidents occur time and again. In addition, this raw material is only available in limited quantities and the oil reserves will be exhausted at some point. (Cf.: Kairos-Institut für Wirkungsforschung, 2019, pp. 12-13)

2.4 Recycling phase

As with the utilisation phase, several points need to be considered in the recycling phase. It depends on the reuse, recycling and disposal of the raw materials.

2.4.1 Recycling

Since the batteries of electric cars contain very rare and sometimes environmentally harmful raw materials, the recycling process focuses on the recovery of these minerals. In the long run, batteries lose more and more of their capacity. When a transaction battery installed in a vehicle has only reached 70 - 80% of its original capacity, it becomes unusable for further use in electric cars. Currently, this performance threshold is reached after about 10 - 15 years. After this time, however, most batteries are not yet disposed of, but are prepared for so-called "second-life" applications. The cells of the battery still have enough power and storage capacity to be used in other applications. After use in electric vehicles, they can still be used, for example, as a house flush, emergency power supply or buffer for rapid charging systems. The service life of the battery can thus be increased by 10 - 25 years. "Second-life" applications protect the environment and save costs and raw materials in the production of new batteries. (Cf.: Kairos - Institut für Wirkungsforschung, 2019, pp. 16-17)

Since there are still relatively few electric cars on the roads, the recycling processes for lithium-ion batteries are not yet fully developed. Extracting the raw materials from the battery cells is very complicated due to their construction. The primary goal is to recover the raw materials cobalt, aluminium and copper. The lithium is only extracted as a mixed secondary product. In order to filter it, an additional step is required, which is more complex and costly.

The more material to be recycled is available on the market, the more efficient the recovery processes become. Electrolytes and various anode coatings could also be filtered out by such processes. As there will be more and more electric cars on the roads in the future, the greater demand for batteries and the associated increase in the price of lithium will have a positive influence on the development and implementation of efficient recycling methods of lithium-ion batteries. (Cf.: Proissl, 2018)

The first recycling company of its kind is located in Antwerp, Belgium. It recycles old cell phone batteries. The process recovers 95% of the processed cobalt, nickel and copper. With lithium, more than 50% can be recovered. In the future, the plant will also recover raw materials from the transaction batteries of electric cars. Recycling is an important point in the life cycle assessment of electric cars. The recycling process requires only a fraction of the energy needed for mining and extraction. (Cf.: Thurn & Scherlofsky, 2019)

2.4.2 Disposal

The increasing demand for electric cars will encourage recycling, but some batteries will always have to be disposed of. The greater the amount of lithium in circulation, the more care will have to be taken with disposal. Lithium is readily soluble in water and can pose enormous health and environmental hazards in groundwater if it is not disposed off properly. Due to the increasing mass of batteries in the waste stream, the risk increases significantly. Just six grams of lithium are enough to kill a healthy adult. (Cf.: Laughlin, 2012, p. 169-171)

2.5 Life cycle assessment - entire life cycle

Taking into account the entire service life of an electric vehicle with the Austrian electricity mix, it turns out that in the entire vehicle cycle up to 90% less greenhouse gases are emitted than in vehicles with combustion engines.

Whereas a gasoline-powered compact car emits approx. 195g CO2-equivalent per passenger kilometer, an electric car emits only about 90g. If the electric car runs on 100% green electricity, only 25g are emitted. It should be noted that the electricity mix of the electricity used plays a decisive role in the ecological balance. The greener the electricity, the fewer emissions result from driving. If a country uses more coal and nuclear power, for example, this has a negative effect on the electricity mix and thus also on the CO2 balance. The electric car becomes more and more efficient as it is used for longer periods of time and, despite the high manufacturing costs, emits on average significantly fewer greenhouse gases than gasoline or diesel. (Cf.: Günsberg & Fucik, 2018, p. 5)

3. PROPULSION

3.1 Electromagnetism

Every electrically operated motor is based on the physical principle that unequal magnetic poles attract and equal poles repel each other. A wide variety of physical principles are applied when building a motor.

When current flows through a conductor, it generates a magnetic field. In a straight conductor, the magnetic field lines form concentric circles. The direction of the field lines depends on the direction of the current. If this conductor is wound several times, a coil is created. If the current remains constant, the magnetic field increases with the number of turns of the coil. An increase in the magnetic field also occurs with the same number of turns and an increase in the current strength. Due to these properties, the magnetic field can be easily controlled.

Just as a magnetic field is generated when a current flows through a conductor, the magnetic net field also leads to a current flow within the conductor. This requires a magnetic field that changes over time and a conductor that can drive current in a closed circuit. In the same way the conductor or coil can move in a magnetic field. The decisive factor is that the magnetic flux changes over time. A temporal change of the flux leads to an induced voltage. This principle is called induction.

An electric motor also needs two magnetic fields. At least one of these magnetic fields must be generated in such a motor by means of a current-carrying conductor. This is necessary to be able to control the magnetic field perfectly. The other field can either be generated in the same way or by a permanent magnet. (Cf.: Schoblick, 2013, p. 181-191)

3.2 Principle of a three-phase motor

Of all the different electric motors, the three-phase motors are among those which are best suited for use in electric cars and which are nowadays also installed in electric cars.
In three-phase motors, torque is generated by the attractive force of the magnets contained in the two motor components, rotor and stator. The moving part is called the rotor and the fixed part the stator. Figure 3 shows the motor components in a recognizable way.

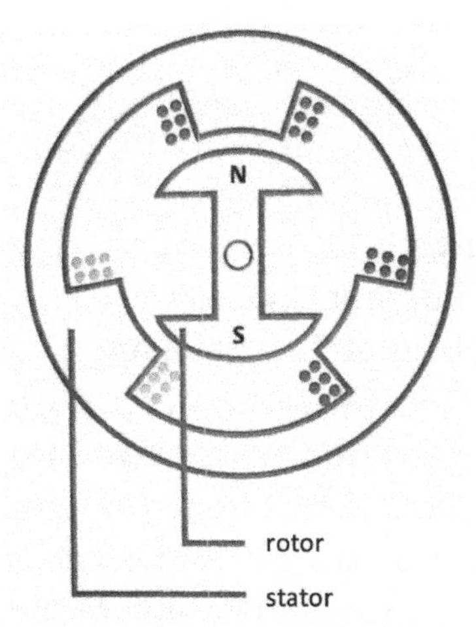

Figure 3, Schematic drawing of the synchronous motor

Three coils, offset by 120°, are supplied with three-phase current. One coil is always supplied with one phase of the rotary current. Three-phase current is an alternating current with three phases, so there is one phase for each coil of the stator. Therefore the phases of the current are also shifted by 120° to each other. This combination creates a magnetic field that circulates with the mains frequency. If the rotor has at least one corresponding counter magnet, it follows this magnetic field. For three-phase motors relevant for electric cars, a distinction can be made between two types. (Cf.: Karle, 2018, p. 63)

3.2.1 Synchronous Motor

Figure 3 shows the principle of a synchronous motor. The coils in the stator generate a rotating field by induced current. The counter magnet (also called rotor) is located on the rotating rotor. The rotating rotor is always synchronous with the rotating magnetic field generated by the stator.
Where the magnetic field of the stator is always generated by current-excited coils, that of the rotor can be generated either by using permanent magnets or coils. (Cf.: Karle, 2018, pp. 63-64)
The speed of synchronous motors is always directly proportional to the frequency of the rotating field generated by the stator. Since the speed is independent of the load on the motor, such a motor is ideal for electric cars. In this case, the speed can be controlled very well via the frequency of the rotating field even in "fluctuating load situations". When starting an electric car with a synchronous motor,

the frequency of the rotating field is lowered to enable starting under load. (Cf.: Schoblick, 2013, p. 221)

- In **permanently energized synchronous motors**, a strong permanent magnet is installed in the rotor. Very strong permanent magnets are required for this purpose, most of which contain rare earths such as neodymium. These elements are not only expensive, but their degradation also causes environmental pollution in the areas of production.

This type of motor offers an advantage in that it achieves good efficiency even at low and medium speeds. The use of an already magnetic rotor means that no slip rings need to be fitted there, which applies voltage to the rotating coils, as with a separately excited synchronous motor.

- Instead of a permanent magnet, an electromagnet is used on the rotor in **externally energized synchronous motors**. As already mentioned, such a motor requires slip rings on the rotor, which serve as sliding contacts and can thus induce current into the coil. These contacts wear out continuously, which means that the motor requires maintenance.

The achieved efficiency and performance level of a permanently magnetised synchronous motor cannot be fully achieved. However, the cheaper production of a separately energized synchronous motor ensures that motors with permanent magnets are rarely used in electric cars. (Cf.: Schoblick, 2013, p. 182, 223-224)

3.2.2 Asynchronous Motor

In asynchronous motors, the rotating magnetic field is also used to generate torque. However, the design of the rotor differs from that of the synchronous motor. It has a much simpler structure. In the simplest case, the rotor of an asynchronous motor consists only of a cage of short-circuited metal bars. Here the metal bars form short-circuited conductor loops.

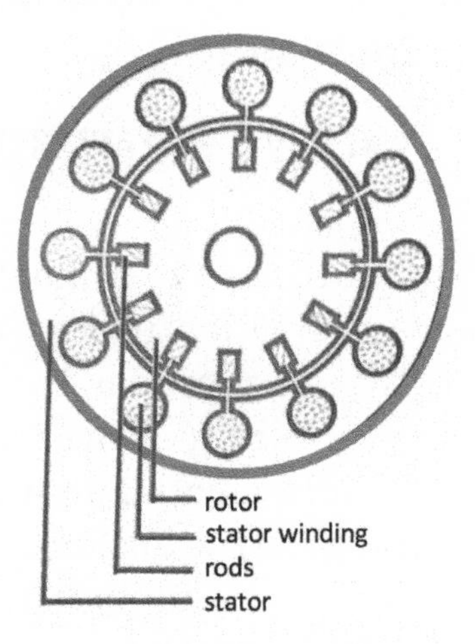

Figure 4, Concept of the asynchronous motor

The asynchronous motor is inexpensive and easy to manufacture. The motor is operated directly from the three-phase mains and offers a high torque over a wide speed range due to its good efficiency. Unlike synchronous motors, the speed here is dependent on the load of the torque. Therefore, the speed of

the motor does not correspond to that of the rotating field and is therefore asynchronous.
Figure 4 shows the structure of an asynchronous motor. By induction of the coils in the stator (stator winding) a current flows in the rods of the rotor (rotor bars). The arrangement of the bars creates a suitable magnetic field in the rotor (rotor). This creates a torque and the rotor follows the rotating field of the stator. (Cf.: Schoblick, 2013, p. 225-228)

3.3 Advantages And Disadvantages Of The Three-Phase Motor

Electric cars have many **advantages** due to the three-phase motors:

- Electric motors have a much simpler design and are smaller in size than a combustion engine with comparable performance. The simpler design of the motor makes it easier to build electric motors with a lot of power. The maintenance of the motor is also not as expensive as with combustion engines. The compact design enables space-saving installation and lower costs. The design makes it easy to switch to reverse gear without having to install a gearbox.
- For technical reasons, the efficiency of an electric motor is significantly better than that of a combustion engine. In the case of

combustion engines, only 260W per kW of power is actually fed to the drive train. For a vehicle with an electric motor, it is between 880W and 960W.

- With combustion engines, chemically stored energy is fed to the engine. In the cylinders, this is converted into thermal energy in order to mechanically set the drive shaft in motion. This process takes some time. The electric motor can generate mechanical torque without detours by directly supplying electrical energy. This allows shorter and more efficient reaction times of the motor.

- In order to maintain optimum torque over a wide speed range, classic cars with combustion engines must be fitted with transmissions, as these can only provide optimum power transmission in a relatively small speed range. An electric car offers a large torque over such a wide speed range, so that the use of a gearbox is not necessary. In practice, only a single-stage reduction gear is installed. (Cf.: Schoblick, 2013, p. 37-43)

- The acceleration of an electric car is better and more direct than that of a vehicle with a combustion engine, thanks to the high torque of the engine at low revs.

- The absence of the need for a gearbox allows very smooth driving with a high level of comfort. The engine is also much quieter in operation and almost silent at low speeds. Especially in conurbations, the noise for the population is significantly reduced.
- Electric motors can theoretically be operated 100% emission-free. If the electricity is 100% renewable, no pollutants are released during the operation of the electric car and during electricity production. Above all, local zero-emission driving offers advantages in conurbations and large cities.
- During a braking process, the braking energy generated can be used. In this process, the engine becomes a generator. This is very simple thanks to the design of the electric motor. This means that energy can be generated by the generator during braking and fed into the battery. This process also relieves the brakes, which means that the brake pads need to be replaced less frequently. The "recuperation" therefore contributes to a more efficient use of energy and vehicle components. (Cf.: Karle, 2018, p. 23-25)

Disadvantages of a three-phase motor:

- In the motor there is little iron and copper loss over time. Iron losses occur when the magnetic poles in the components change. These losses occur continuously and are load-independent. Copper losses are caused by the heating of the coils in the motor and are load-dependent. The coils only heat up when current is supplied.
- The use of permanent magnets in permanently energized synchronous motors requires rare earths. The element neodymium, which is required for strong permanent magnets, is particularly noteworthy here. The raw material is only available in limited quantities and the mining process is associated with adverse effects on the environment.
- The noiseless operation of the engine can cause pedestrians and passengers to perceive nearby vehicle traffic less well and thus notice it only later or not at all.
- Some disadvantages of electric cars are not blamed on the engine, but on the storage battery.

4. STORAGE SYSTEMS

4.1 Battery as energy storage

The most common storage method currently used in electric cars is rechargeable batteries. These make it possible to "refuel" the car directly with electricity and store it.
Almost all batteries installed in electric cars are lithium-ion batteries. Here, many small lithium base cells are connected together to serve as a large vehicle battery, also called transaction battery. Lithium is used because of its chemical properties.

4.1.1 Lithium-Ion Batteries

Lithium forms slightly positively charged lithium ions, so-called cations. In a rechargeable battery, these cations act as charge carriers that transport the electrical charge between the two electrodes. The anode is made of graphite, as this has a relatively high energy density due to its layer struc-

ture. The cathode consists of lithium cobalt oxide, whereby metals such as nickel or manganese can be used instead of cobalt. Meanwhile, there are many different ways to produce cathodes with different materials. Between the electrodes there is an organic electrolyte. This electrolyte must be conductive and able to withstand a cell voltage of about four volts. (Cf.: Bresser, 2018)
During the charging process, the lithium cations are embedded in the graphite anode. When discharging, they go back into solution and migrate to the cathode. During each charging and discharging cycle, the lithium cations always change to the other electrode and either take up an electron or give one away. (Cf.: Ducci & Oetken, 2018)

Although there are already many good approaches to produce lithium-ion batteries with even higher energy capacity, many of these concepts for new batteries are still in the starting blocks. One example would be to replace the graphite anode in conventional lithium-ion batteries with silicon. Such a change would enable a battery capacity ten times greater. However, silicon expands by up to 300% when it absorbs the lithium ions. This increase in volume leads to a fragmentation of the silicon anode after only a short time. (Cf.: McDowell, 2018)

4.1.2 Advantages And Disadvantages Of Batteries

Advantages of using batteries in electric vehicles, especially lithium-ion batteries

- Compared to other battery types, lithium-ion batteries offer a high energy and power density. The discharge, which increases with increasing time, is also relatively low.
- Without further consequences on the battery life, lithium-ion batteries can be charged from any charge state without having to be completely discharged beforehand.
- Thanks to the high energy density of the graphite, these batteries have a high efficiency. Other types of batteries perform worse due to the use of other elements.
- If an electric vehicle has a rechargeable battery and is therefore "fueled" with electricity, it is theoretically possible to operate the vehicle completely with environmentally friendly green electricity. Even though emissions are produced during the generation of electricity, the electric car does not emit any local pollutants.
- It is much cheaper to charge the vehicle with electricity than to refuel it with fossil fuels.
- A heavy transaction battery offers the ad-

vantage that it is built into the electric car and provides a lot of stability while driving. The car's centre of gravity is close to the ground, which has a positive effect on driving behaviour. This significantly reduces the probability of rollover even in the event of a collision.

Disadvantages of the battery:

- The life span of the battery is limited, because with each charging cycle the capacity of the battery decreases a little bit. Lithium-ion-batteries are far ahead of many other types of batteries, but depending on the usage, after approximately 10-15 years, they only reach approximately 70-80% of their former capacity and are not usable anymore for the usage in electric cars.
- Lithium-ion batteries must not be overcharged or deeply discharged. To prevent this process, the batteries must be equipped with additional protective electronics.
- Batteries can only be charged with a limited charging current. This means that electric cars sometimes have long charging times. Especially in comparison to the burners, which can be fully charged within a very short time.

- In order to work optimally, lithium-ion batteries must be kept in a temperature range between 18°C and 25°C. This means that the transaction battery needs to be air-conditioned, which requires additional energy. This can reduce the range of an electric car on particularly hot or cold days.
- Although lithium-ion batteries have a high energy density compared to other types of batteries, this density is very small in direct comparison with the burners. Where approx. 12 kWh of energy is chemically stored in one kilogram of petrol, the energy quantity of a lithium-ion cell of the same weight only amounts to about 0.13 kWh.
- When storing electricity in the battery, a small amount of energy is always lost in the long run. With a petrol/diesel tank the efficiency is 100%.
- The many battery cells required make up quite a bit of the weight. In order to be able to transport this additional weight, the built-in motor requires correspondingly more power.
- As already described in point **2.2 Battery production**, many rare earths are needed for a vehicle battery. A lot of C02 is released during the production of such a battery.

The mining of these raw materials is also harmful to the environment.

Especially the required lithium is very toxic. The mining has a massive impact on the ecosystem. Both the recycling and the disposal of large quantities of lithium pose a problem.
(Cf.: Karle, 2018, p. 43-45, 78-81)

- The battery of an electric car is very difficult to extinguish in case of fire. The vehicles usually burn out completely without the fire brigade's extinguishing measures being of great benefit. In the case of an apparently extinguished electric car, there may still be short circuits in the battery cells some time later. (Cf.: Viehmann, 2019)

4.2 Hydrogen as energy storage

If you want to drive electrically with hydrogen, you have to face a fundamental problem: How does hydrogen become electrical energy for the engine of an electric car?

4.2.1 Hydrogen Fuel Cell

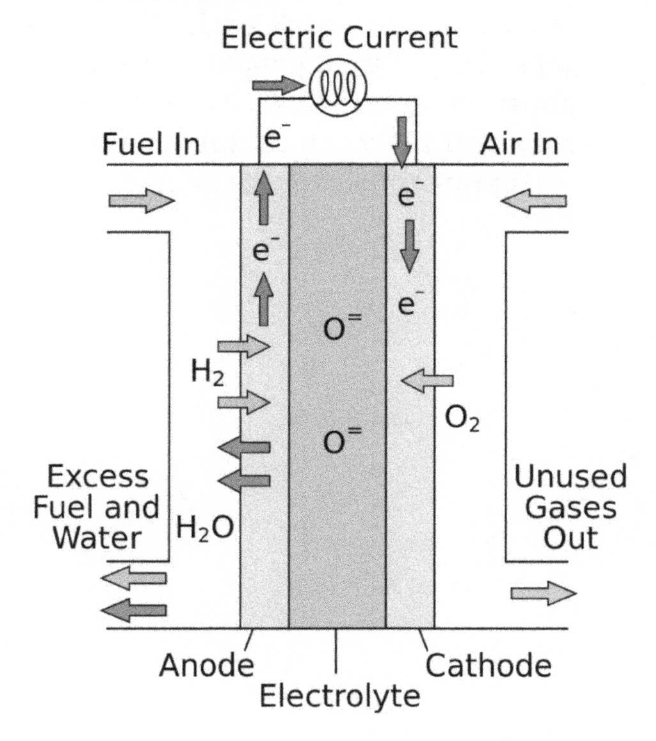

Figure 5, How a hydrogen fuel cell works

Fuel cells are the solution to this problem. Figure 5 shows the structure of a hydrogen fuel cell. In an electric car, the required hydrogen (H) is filled up and oxygen (O) is supplied from the air. To generate electrical energy, the hydrogen is split at the anode into an electron and a positive hydrogen ion (H+), a cation. A catalyst is required to start this reaction. In most cases this consists of platinum. The hydrogen cations (shown here in red) now move through the electrolyte membrane. Since the electrons cannot pass through this membrane, they take a detour. Here the electrons can be used to generate energy. After the hydrogen cations have reached the oxidant at the cathode (shown here in blue), they combine with the oxygen, which has previously taken up an electron, to form water (H2O). (Cf.: Randelhoff, 2019)

- **Reaction at the anode:** 2H2 → 4H+ + 4e-

  Hydrogen atoms (H) split into H-cations and electrons.

- **Reaction at the cathode:** O2 + 4H+ + 4e- → 2H2O

  The added oxygen (O) combines with the H cations and electrons to
  water molecules (H2O).

Since a single fuel cell can produce little energy, several cells are connected together to supply the vehicle with sufficient electrical energy, similar to transaction batteries. Although this process of energy generation produces pure water as a "waste product", the practical application of fuel cells in the automotive sector is difficult. Production and storage of hydrogen are expensive and complex. (Cf.: Schoblick, 2013, p. 301 - 303)

4.2.2 Advantages And Disadvantages Of Hydrogen

This technology offers the following **advantages** as a storage for electric cars:

- The high-pressure tank for the hydrogen is relatively compact and offers a good range. The gas can therefore be compressed and stored.
- Only a small battery is needed. This is only used to supply starting current and to recover energy during braking.
- Filling the tank with hydrogen takes only a few minutes.
- Hydrogen can be produced 100% green, namely if it is produced by electrolysis of water and the energy required for this is purely ecological (e.g. comes from a local photovoltaic system).
- Locally, a car with a fuel cell emits no pollutants.

- The fuel cell itself has a very high efficiency. This remains constant even under partial load of the fuel cell.

Disadvantages of the hydrogen fuel cell:

- Despite its potential to be good for the environment, about 90% of hydrogen is derived from fossil fuels.
- The production of hydrogen by electrolysis and the subsequent compression or liquefaction is very lossy. Ultimately, the efficiency is only around 26%. For an electric car with a rechargeable battery it is almost 70%.
- To refuel cars with hydrogen fuel cells, a network of filling stations is required. At the moment there are only about 60 such filling stations in Germany. There are about 15,000 filling stations for internal combustion engines and about 9,000 charging points for electric vehicles with rechargeable batteries. Further expansion of this filling station network would require immense costs.
- The volatile hydrogen only becomes liquid at -253 °C. It is therefore very difficult and costly to transport the hydrogen from the production plant to the filling station.

- Currently, only a small number of hydrogen cars are produced. This makes the technology very expensive. It would only change if larger quantities of these cars were built.

- Hydrogen is explosive and can be dangerous if used incorrectly. However, this often leads to the assumption that this gas is very dangerous and difficult to handle. With today's standards, an incident is very unlikely. (Cf.: Stüber, 2019)

4.3 Methanol as energy storage

Electric vehicles can be powered not only by hydrogen, but also by the proper use of methanol. Methanol is an alcohol, which can be produced either synthetically or by biomass reactors. The question now is how such an alcohol can power an electric car.

4.3.1 Direct Methanol Fuel Cell

The solution to the problem lies, as with hydrogen, in the fuel cell technology. The principle is very similar to the hydrogen fuel cell. Methanol (CH4O) is split together with water (H2O) at the anode by a catalyst. The hydrogen of the split molecules gives off its electrons. The positive hydrogen ions now migrate through the electrolyte membrane to the cathode, where the hydrogen cations then combine with the oxidant oxygen and the electrons to form water. This reaction leaves CO_2 at the anode and water at the cathode.

- Reaction at the anode: $2CH_4O + 2H_2O \rightarrow 2CO_2 + 12H^+ + 12e^-$

Methanol molecules (CH_4O) and water molecules (H_2O) are split up. The hydrogen (H) thus separated splits into H-cations and electrons. The remaining atoms combine to form carbon dioxide (CO_2).

- Reaction at the cathode: $3O_2 + 12H^+ + 12e^- \rightarrow 6H_2O$

The oxygen (O_2) supplied to the cathode reacts with the H-cations and electrons to form water (H_2O) (Cf.: Schoblick, 2013, p. 303)

4.3.2 Advantages And Disadvantages Of Methanol

A methanol fuel cell offers these **advantages**:

- The system consisting of fuel cell and methanol tank has ten times the power density of a lithium-ion battery. This means that a lot of energy can be stored in a small space and weight and space can be saved.

- As with the hydrogen fuel cell, only a very small battery is needed compared to the electric car with battery. (Cf.: Purtul, 2009)
- Methanol can be easily transported. The liquid alcohol can easily be used with existing infrastructure. It can be filled up at petrol or diesel filling stations. The existing network of filling stations is used for this purpose.
- Methanol is easy to produce. The alcohol can be produced in biomass reactors. This method of methanol production is very environmentally friendly. Only as much CO2 is released into the atmosphere by the fuel cell as was extracted from it during plant growth. (Cf.: Schoblick, 2013, p. 303-304)
- The efficiency remains constant even under partial load, i.e. when the fuel cell is not operating at maximum power. (Cf.: Paschotta, 2019)

Disadvantages of the methanol fuel cell:

- It's a little more complicated to split methanol than hydrogen. This means that the process requires larger quantities of the catalyst platinum. Platinum is a rather rare and above all expensive material, which has a negative effect on the costs of the fuel cell.

- Filling up with methanol is more expensive in direct comparison with the price of electricity. The bigger problem, however, is the lack of demand for methanol fuel cell cars. Similar to hydrogen, there are only a few such cars on the roads. The small amount of these cars makes them very expensive. (Cf.: Purtul, 2009)
- Methanol is toxic and not edible alcohol. Possible abuse must be taken into account.
- During the production of energy, carbon dioxide is produced locally in the fuel cell. Hydrogen produces no local emissions. (Cf.: Schoblick, 2013, p. 303)
- The energy emitted by a fuel cell cannot be retrieved as quickly as that from a battery.
- The achievable efficiency of a direct methanol fuel cell is much lower than that of a hydrogen fuel cell. (Cf.: Paschotta, 2019)

5. FUTURE VIABILITY

5.1 Targets for CO2 reduction

In December 2008, the EU Council and the EU Parliament agreed on a regulation that limits the CO2 emissions of new passenger cars. Since 2020, the limit value for carbon dioxide emissions has been binding. This means that 95g CO2/km may no longer be exceeded as the limit value for a new car. This value corresponds to a diesel consumption of 3.61l/100km. In order to comply with these limits for the entire vehicle fleet, car manufacturers have been relying more and more on plug-in hybrids and electric vehicles in recent years. According to the EU, electric cars with 0g CO2 emissions are included in the calculation of the fleet value. Without electric cars, such a CO2 limit would hardly be possible in the EU. (Cf.: Schoblick, 2013, p. 27)

5.2 Forecasts and prospects

This EU regulation, combined with the ever-increasing consumer demand, ensures that car manufacturers are upgrading their fleets. At the end of 2018, there were only 60 different electric cars, plug-in hybrids and vehicles with fuel cells in Europe. It is estimated that by 2025 there will be over 330 different models. The large car manufacturers are thus practically forced to increasingly deal with new and more environmentally friendly alternatives. These technologies are becoming more and more affordable and cost-effective due to the growing market. Already now there are various monetary subsidies for the purchase of electric cars. (Cf.: Smartics, 2019)

Intensive research is already underway into new and improved battery technologies. The future market prospects and higher demand mean that research is constantly advancing. At every point, efforts are being made to improve batteries, be it the charging time by using different coatings or the capacity with the help of new electrolytes and electrode materials. Battery technologies have already made immense progress in recent years. (Cf.: Schlak, 2019)

6. CONCLUSION

As the work process progressed, it became increasingly clear to me how many different factors are involved in the life cycle assessment of an electric car. Because of their large transaction battery, electric cars already have high CO_2 emissions at the beginning of their life cycle, i.e. exactly what one would actually like to reduce by producing such vehicles. The raw materials required for their production also bring with them some serious environmental effects. But once such a vehicle is on the road, it can score points for its quiet and efficient operation. No oil has to be pumped out of the ground to power a car with an electric motor. Renewable energy sources give the electric car the lead in CO_2 emissions. With electricity from wind, water or solar energy, electric cars can be operated in an environmentally friendly way with minimal emissions. There are therefore very few emissions during driving, unlike with combustion engines, which emit more and more air pollutants with increasing lifetime. If a country's energy comes mainly from fossil sources, such as coal, for example, electric cars and combustion engines are almost equal in terms of emissions. Even the recycling of old batteries is currently still complicated and difficult.

However, the relevance of the electric car cannot be assessed by one criterion. A big advantage of the electric car is its engine. The electric motors used are far more efficient and climate-friendly than combustion engines. Their compact design and high efficiency make them almost ideal for use in vehicles.
The current required for the motor can be stored in various ways. As mentioned before, there are rechargeable batteries which are a great burden on the environment during production. Lithium-ion batteries are energy-intensive in production and require a large amount of rare raw materials, the mining of which causes many problems. However, electricity can also be generated directly in the vehicle with the help of fuel cells. In this case, the vehicle would not need to be charged directly with electrical energy, but can be refuelled with hydrogen or methanol. However, these technologies are less common than battery-powered electric cars.
In the future, thanks to their efficient driving behavior, electric cars can do much to reduce emissions of pollutants into our environment and atmosphere. Advancing research is making electric cars increasingly suitable for the mass market, be it with new battery technologies or fuel cells. The most important point, however, is that energy production, whether for use in batteries or for the production of hydrogen or methanol, is becoming increasingly renewable. Running an E-PKW on energy from fossil fuels contradicts the intention to

be increasingly independent of oil and natural gas. Electric cars have great potential to relieve our environment. However, this can only happen if our energy infrastructure becomes more sustainable, more conscious and above all more environmentally friendly.

REFERENCES

Printed literature:

Bresser, D. (Mai 2018). Energiespeicher für eine elektromobile Gesellschaft. *Spektrum der Wissenschaft*, P.21-25. Heidelberg, Deutschland: Spektrum der Wissenschaft Verlagsgesellschaft.

Buchert, M. (Juli 2018). Elektromobilität: Ausgebremst durch Rohstoffmangel? *Spektrum der Wissenschaft*, P.46-51. Heidelberg, Deutschland: Spektrum der Wissenschaft Verlagsgesellschaft.

Ducci, M., & Oetken, M. (September 2018). Wie Lithiumakkus funktionieren. *Spektrum der Wissenschaft Kompakt, Energiepseicher: Akkus und Batterien der Zukunft*, P.14-22. Heidelberg, Deutschland: Spektrum der Wissenschaft Verlagsgesellschaft.

Günsberg, G., & Fucik, J. (Januar 2018). *Faktencheck E-Mobilität: Was das Elektroauto wirklich bringt.* Wien, Österreich: Klima - und Energiefonds, VCÖ - Mobilität mit Zukunft.

Kairos - Institut für Wirkungsforschung. (Februar 2019). *Umwelteffekte von Elektromobilität.*
Bregenz, Österreich: Amt der Vorarlberger Landesregierung.

Karle, A. (2018). *Elektromobilität: Grundlagen und Praxis.* München, Deutschland: Carl Hanser Verlag.

Laughlin, R. B. (2012). *Der Letzte macht das Licht aus: Die Zukunft der Energie.* München, Deutschland: Piper Verlag.

McDowell, M. T. (September 2018). Neue Elektroden für Lithiumakkus. *Spektrum der Wissenschaft Kompakt, Energiepseicher: Akkus und Batterien der Zukunft*, P.24-26. Heidelberg, Deutschland: Spektrum der Wissenschaft Verlagsgesellschaft.

Schoblick, R. (2013). *Antriebe vn Elektroautos in der Praxis: Motoren, Batterietechnik, Leistungstechnick.* Haar bei München, Deutschland: Franzis Verlag.

Online sources:

Lauerer, M. (16. Oktober 2018). *Lithium: Abbau und Gewinnung - Umweltgefahren der Lithiumförderung.* Retrieved 25. August 2019 from Edison Handelsblatt: https://edison.handelsblatt.com/erklaeren/lithium-abbau-und-gewinnung-umweltgefahren-der-lithiumfoerderung/23140064.html

Paschotta, R. (4. Dezember 2019). *RP-Energie-Lexikon - Brennstoffzelle, Wasserstoff, Methan, Methanol, Brennstoffe, Niedertemperatur, Hochtemperatur, Andwendugen, Elektroauto.* Retrieved 24. Januar 2020 from Energie-Lexikon: https://www.energie-lexikon.info/brennstoffzelle.html

Proissl, A. (11. Dezember 2018). *E-Atuo-Akkus: Recycling für tickende Umweltbomben.* Retrieved 25. August 2019 from Trend: https://www.trend.at/branchen/auto-mobilitaet/elektroauto-akku-recycling-10506019

Purtul, G. (26. November 2009). *Technik im Trend: Mobile Minikraftwerke.* Retrieved 20. Januar 2020

from Zeit: https://www.zeit.de/2009/49/Methanol-Brennstoffzellen/komplettansicht

Randelhoff, M. (7. Juli 2019). *Wie funktioniert ein Brennstoffzellenfahrzeug?* Retrieved 27. Dezember 2019 from Zukunft Mobilität: https://www.zukunft-mobilitaet.net/77641/zukunft-des-automobils/elektromobilitaet/wie-funktioniert-ein-brennstoffzellenfahrzeug-technik-kritik-bewertung/

Schlak, M. (30. Oktober 2019). *Akkus: Neue Hochenergie-Batterie lässt sich in zehn Minuten laden.* Retrieved 5. Februar 2020 from Spiegel: https://www.spiegel.de/wissenschaft/technik/akkus-neue-hochenergie-batterie-laesst-sich-in-zehn-minuten-laden-a-1294119.html

Smartics. (5. August 2019). *Elektrifizierende Zukunft - Marktprognose Elektroautos bis 2015.* Retrieved 5. Februar 2020 from Smartics: https://smatrics.com/news/marktprognose-elektroautos-bis-2025

Stüber, J. (12. Februar 2019). *Brennstoffzelle: Was bringt die Wasserstoffzelle fürs Auto?* Retrieved 28. Dezember 2019 from Welt: https://www.welt.de/wirtschaft/gruenderszene/article188631963/Brennstoffzelle-Was-bringt-die-Wasserstofftechnologie-fuers-Auto.html

Thurn, V., & Scherlofsky, N. (Produzenten). (2019). *Elektroautos - Schattenseiten eines Booms* [Dokumentation]. Retrieved 10. Januar 2020 from https://tvthek.orf.at/profile/Weltjournal/5298609/WELT-journal-Elektroautos-Schattenseiten-eines-Booms/14037415/WELTjournal-Elektroautos-Schattenseiten-eines-Booms/14621091

Viehmann, S. (30. Juli 2019). *Tesla Model-S brennt in Ratingen ab: Hat Elektro-Star unsichere Akkus verbaut?* Re-

trieved 5. Februar 2020 from Focus: https://www.focus.de/auto/elektroauto/thermal-run-away-im-elektroauto-tesla-kennt-die-brandgefahr-seiner-akkus-seit-10-jahren_id_10620636.html

TABLE OF FIGURES

CPSIA information can be obtained
at www.ICGtesting.com
Printed in the USA
LVHW091330240520
656464LV00004B/1137